T0159079

All the Tea in China

All the Tea in China

History, Methods and Musings

By Wang Jian
Translated by Tony Blishen

Better Link Press

Copyright © 2013 Shanghai Press and Publishing Development Co., Ltd.

All rights reserved. Unauthorized reproduction, in any manner, is prohibited.

This book is edited and designed by the Editorial Committee of *Cultural China* series.

Managing Directors: Wang Youbu, Xu Naiqing
Editorial Director: Wu Ying
Editors: Zhang Yicong, Yang Xiaohe

Compiled by Wang Jian
Translated by Tony Blishen
Paintings by Zhang Lifeng
Photographs by Wang Jian, Getty Images, Quanjing

Designer: Wang Wei
Cover Image: Getty Images

ISBN: 978-1-60220-137-8

Address any comments about *All the Tea in China: History, Methods and Musings* to:

Better Link Press
99 Park Ave
New York, NY 10016
USA

or

Shanghai Press and Publishing Development Co., Ltd.
F 7 Donghu Road, Shanghai, China (200031)
Email: comments_betterlinkpress@hotmail.com

Printed in China by Shenzhen Donnelley Printing Co., Ltd.

3 5 7 9 10 8 6 4 2

On page 1

There is a kind of contentment called "leisure" and a kind of happiness known as "savoring."

On page 2

What is the flavor of life? Look back on all its twists and turns and it is without incident.

Contents

Preface

Tea not only quenches thirst and cleanses the palate, it also raises the spirits, clears the mind and keeps the body fit. It is the world's most welcome beverage as well a unique cultural vehicle.

China has a long history of tea drinking. In the beginning, because of its detoxicant and thirst-quenching properties it occupied a major place in the physical world of Chinese food and drink; later it combined with poetry and art and tea tasting gradually permeated the spiritual and cultural world, forming and developing a unique Chinese tea culture.

Today, tea is still a major and indispensable factor of Chinese life.

Consequently we have produced this book in the hope that we can share the fascination of tea and bring the reader to a fuller understanding, greater intimacy and love of tea.

In this book we quote proverbs, poetry and stories about tea from both past and present, dynasty by dynasty and period by period. Through this the reader will be able to experience the

◀ Do not envy the full fruit of others, take one's ease beneath one's own trellis.

spiritual appreciation and joy that tea has brought to the Chinese people at different places and times and thus realize the spiritual context that tea has come to acquire.

At the same time, the book's illustrations and text present the origins, varieties, manufacturing processes and etiquette of tea drinking so that the reader can gain practical information about what tea is, how it is drunk and tasted and understand how the tea culture came about.

It should be explained that this book is both systematic and yet diffuse. It is systematic in its arrangement of quotations, proverbs and poetry by historical period in order to allow prominence to the characteristics of each dynasty. It is diffuse to the extent that in order to meet the demands of the tempo of life, each illustration and piece of information stands independent of the rest so that the reader can read from any point or page without fear that the logic of the narrative has been disturbed.

Nevertheless, the book as a whole displays a multi-faceted, three-dimensional image of Chinese tea culture.

Make a pot of tea, open the book and let us together enjoy the gentle delight, tranquility and spiritual appreciation of Chinese tea.

► Without the sour taste of an unripe persimmon who can know the beauty of its matured flavor?

柿熟時才
會記起青澀
或許青澀的
味道最美
沒有生澀的
味道誰知道
柿熟的甘美
是如此的美好
秋峰

香茗
勝小酒
丙戌秋

Pre-Qin Period

(21st Century BC – 221 BC)

The history of mankind's discovery and use of tea can be traced back four to five thousand years. In the beginning man used the fresh leaves of the tea tree as a vegetable food. Later he discovered its medicinal properties and used it to preserve health.

◀ Fragrant tea conquers wine.

When the mind is clear one can sip tea, when the spirit is at ease one can talk of aspiration.

—Ancient Chinese Saying

▶ Ancient Tea Tree

There are many references to wild tea trees in ancient Chinese texts. Numerous wild tea trees of considerable age are still to be found in the mountainous areas of southwest China.

In 1961, a 1,700 year old wild tea tree with a height of 32.12 meters and a diameter of 1.21 meters was discovered in the primeval forests of the Dahei Mountain of Menghai county in Yunnan province.

In 1996, an ancient tea forest of an area of approximately 280 hectares was discovered in Jiujia township, Zhenyuan county, Yunnan province. This is the world's largest stand of wild tea trees.

The right picture shows the ancient tea tree discovered in 1996 in Zhenyuan county, Yunnan province.

▲ *Shennong's Materia Medica*

According to ancient Chinese texts tea leaves had the ability to relieve the symptoms of poisoning. This was first discovered by Shennong in the pre-historic period. *Shennong's Materia Medica* was compiled during the Eastern Han period (25 – 220). It is the earliest surviving Chinese pharmacological text.

Shennong tasted every plant and encountered 72 poisons in a day but dispersed them with tea.

—*Shennong's Materia Medica*

Shennong, also known as the Yan Emperor (literally flame emperor), was the sun god of Chinese pre-historic mythology and the inventor of agriculture who taught mankind how to till the earth.

What matters it if there is no wine for a guest? As long as there is real feeling, to chat and sip tea the while is true contentment.

—*Ancient Chinese Saying*

▶ Tea Producing Areas

From the national level, Chinese tea producing areas are divided into four, and the famous teas from each area are represented by:

Regions south of the Yangtze River: Xihu Longjing (Dragon Well), Huangshan Maofeng (Feather Mountain), and Biluochun (Green Conch Spring).

Regions north of the Yangtze River: Lu'an Guapian (Melon Strip) and Xinyang Maojian (Feathertip).

South China: Tieguanyin (Iron Buddha) and Wuyi Dahongpao (Crimson Gown).

Southwest China: Pu'er tea and Dianhong tea.

▲ Different Kinds of Teas

China tea is divided into green, red, Wulong (Oolong), white, yellow and black according to the method of manufacture.

Green and yellow tea is unfermented. The color of green tea after infusion is a verdant green and yellow tea a yellowish color.

White tea is slightly fermented and pale after infusion.

Wulong tea is semi-fermented and displays a golden color after infusion.

Red tea is fermented and has mainly a reddish yellow color after infusion.

Black tea is extra fermented and orange-yellow after infusion.

Who says tea is bitter? It's as sweet as shepherd's purse.

—*The Classic of Poetry*

The Classic of Poetry or *Shi Jing* was compiled during the Spring and Autumm period (770 – 476 BC) and is the first collection of poetry and song in China. It comprises a total of 305 folk songs and pieces of court and temple music dating from the 500 years between the 11th and 6th centuries BC.

Tea clear the mind but wine confuses the spirit.

—*Ancient Chinese Saying*

▶ Producing Green Tea

Green tea is the most produced and most widely drunk tea. The color of the dry tea, of the tea after infusion and of the actual leaf is mainly green hence the name "green tea." Examples of famous teas: Xihu Longjing (Dragon Well), Biluochun (Green Conch Spring).

The art of tea-making:

Step one: *shaqing* (fixing the tea by treating it at a high temperature to regulate fermentation, thus allowing the leaves to retain their original color whilst at the same time reducing the moisture content and softening them for further processing). Fixing by steam is known as "*zhengqing*." Fixing by stir-drying is known as "*chaoqing*." Fixing by baking is known as "*hongqing*." Fixing by sun-drying is known as "*shaiqing*."

Step two: maceration or kneading.

Step three: drying (stir-drying, baking, sun-drying).

一樹秋聲
戊子秋

Qin and Han Period

(221 BC – AD 220)

This period marks the emergence of the tea culture, tea is already in production and has reached the lower reaches of the Yangtze and the Zhejiang littoral by way of the state of Jin Chu (present-day Hubei province) from the states of Ba and Shu (present-day Sichuan province). References to the production of tea and its medicinal properties appear in literature.

◀ Cleanse the heart of earthly desires and it will be difficult to exhaust joy.

Drinking tea over time strengthens thought.

—*Hua Tuo: Dissertation on Foods*

Hua Tuo (c.141 – 208) was a famous early doctor and inventor of the anaesthetic *mafeisan* (believed to be a cannabis derivative) which allowed the performance of surgery on an anaesthetised patient, the world's earliest recorded instance of surgery after a general anaesthetic.

► Huangshan Maofeng (Feather Mountain)

One of China's most famous teas.

Type: Green tea.

Area: Mt. Huangshan in Anhui province.

Characteristics: Slightly rolled and with a hint of yellow in the green. Special quality Maofeng is picked and processed either side of the Qingming Festival. It is of the highest quality, its leaves shaped like a sparrow's tongue, ivory colored and scented like an orchid with the fragrance remaining after the tea has cooled.

Best drunk from a glass.

At Jingzhou in Hubei and Bazhou in Sichuan tea leaves are picked and made into blocks…which when imbibed cure the effects of wine and prevent sleepiness.

—*Guangya*

The *Guangya* was compiled during the period of the Three Kingdoms (220 – 280) and is one of the earliest Chinese dictionary.

◀ Biluochun (Green Conch Spring)

One of China's most famous teas.

Type: Green tea.

Area: Dongting Mountain, Lake Taihu, Suzhou, Jiangsu province.

Characteristics: Tightly twisted leaf resembling a conch shell (*luo*). Over-powering fragrance and originally called "stop-you-in-your tracks scented" by local people (Suzhou dialect for pungent).

Picking: Starting from approximately Spring Equinox (21 March – 4 April), ending approximately Grain Rain (20 April – 4 May). The best and most highly valued crop is picked and processed between the Spring Equinox and the Qingming Festival (roughly late March to early April).

Best drunk from a glass.

A monk can sustain life on three *mou* (1 *mou* equals 666 square meters) of tea garden and a fisherman live by his rod alone.

—*Chinese Proverb*

Infusing Green Tea

A clear glass bowl is best for infusing green tea as it is possible to observe and enjoy the whole process of the unfolding and sinking of the tea after it has been placed in the water. The best temperature for infusing green tea is approximately 85 degrees centigrade.

The three usual methods of infusing green tea are:

All water first. Fill the cup with boiling water and then add enough tea leaves. This is the method used for Xihu Longjing (Dragon Well), Biluochun (Green Conch Spring), and Xinyang Maojian (Feathertip) teas.

Some water first. Fill the cup one third full of boiling water, add enough tea leaves and then top up with boiling water. Used for Huangshan Maofeng (Feather Mountain) and Lu'an Guapian (Melon Strip) teas.

Tea first. First put the tea leaves in the cup and then add boiling water. This method can be used for infusing stir-dry fixed green tea during the winter.

Drinking tea enlarges the spirit and delights the mind.

—*Shennong's Classic of Food*

Shennong's Classic of Food was compiled during the Western Han period (202 BC – 25 AD) and is the earliest Chinese recorded work about food. The complete text has been lost.

◀ Red Tea

Red tea is the second major variety of Chinese tea. The many oxidants in the leaf produce a red pigmentation which accumulates in the leaf and dissolves in water causing the typical red appearance of both leaf and hot water, hence the name. Examples of famous teas: Keemun tea, Dianhong.

It is produced as follows: Step one: dessication, that is spreading the fresh leaves in the open air and sun to reduce the moisture content. Step two: kneading. Step three: fermentation. Step four: drying.

Wei, Jin and Northern-Southern Period

(220 – 581)

China's tea culture blossoms. The production of tea and the habit of tea drinking gradually spread throughout south of the Yangtze. Tea is used to greet guests and in sacrifices. Tea is also used by the literati as a partner in the expressive recitation of poetry.

◀ A bowl of fragrant tea benefits poetry and the spirit may be seen in two or three bamboos.

In the final years of the Three Kingdoms when Sun Hao was ruler of the state of Wu he stipulated that at every banquet each minister should drink at least seven liters of wine. The minister Wei Yao was well regarded by Sun Hao on account of his learning but could not hold his liquor. Sun Hao thereupon stealthily substituted tea for his wine thus sparing Wei Yao the embarrassment of appearing the worse for drink in public.

—*Chen Shou: Record of the Three Kingdoms*

Chen Shou (233 – 297) was a historian. His *Record of the Three Kingdoms* is a biographical record of the sixty years between 220 and 280. The story of Sun Hao and Wei Yao is the origin of the Chinese custom of "substituting tea for wine."

▶ Keemun tea

One of China's historically famous teas, known with Indian Darjeeling and Sri Lankan Uva seasonal teas as one of the world's three most aromatic teas.

Type: Red tea

Area: Qimen in Anhui province

Characteristics: Fragrant tightly curled leaves, jet black in color with a gray sheen. It won a gold medal at the 1915 Panama International Exhibition of Local Produce.

This picture shows the beautiful villages and mountains of Qimen, Anhui province.

Busy for renown, busy for profit, but steal time for leisure from busyness by drinking a bowl of tea. Labor in mind and suffer, labor in body and suffer, but sneak joy from suffering by drinking a jar of wine.

—*Chinese Proverb*

◀ Shallow for Tea Full for Wine

A custom of the Chinese tea and wine culture. When pouring tea for a guest the cup is filled seven to eights parts full. When pouring wine the cup is filled to the brim. The reason being that tea is drunk hot and a full cup is difficult to handle. This custom reflects the concern for people that lies at the heart of the Chinese culture of tea.

Softly closing the chamber door,
Alone I await you in the empty room.
You never came.
Disappointed, I drink solitary tea.

—*Wang Wei: An Occasional Poem*

Wang Wei (415 – 443), famous painter and literati.

▶ Tieguanyin (Iron Buddha Tea)

One of China's historically famous teas.

Type: Wulong (Oolong) Tea

Area: Anxi in Fujian province

Characteristics: Leaves tightly rolled into granules. Infusion produces a light natural orchid-like fragrance with water of a golden yellow.

Picking and processing: Can be picked all year round. Spring tea is picked and processed from 20 April to 20 May, summer tea from 21 June to 22 July and autumn tea from 7 August to 22 October. Autumn tea is the best.

Who says that bitter tea won't dispel gloom?
Like a water chestnut its sweetness rids the heart of anxiety.

—*Chinese Poem*

◀ Tea-set for Gongfu Tea

Drinking Wulong (Oolong) tea is often known as taking "Gongfu tea."

The tea-set consists of the four jewels of tea-making: a stove, an earthenware teapot with a capacity around 200 milliliters, an unglazed pottery jug and a tea bowl with a capacity around 40 milliliters (normally a set varies between three to five pieces).

Best infusion temperature: Approximately 100 degrees centigrade.

Sui and Tang Period

(581 – 907)

During the Sui and Tang period, and particularly from the Mid-Tang dynasty onwards tea gradually extends from the court, the nobility and the literati to the lower orders of society and the habit of drinking tea spreads throughout the country.

In this period also, the tea industry begins to produce revenue; a department for the production and provision of tea to the imperial household is set up and the Buddhist tea ceremony begins to flourish. From having the function of a medicine or a physical beverage tea becomes a form of culture, a means of cultivation and a realm of its own.

◀ At leisure and seated alone, one may savor the tea, appreciate the teapot and ponder life.

Take a mouthful of tea and gloom is swept away. Take another mouthful and feelings are clarified and the spirit rejoices. Take yet another mouthful and the heart is purified. In this way how can one still be disturbed by anxiety?

—*Jiao Ran: A Song of Tea*

Jiao Ran, a Buddhist monk of the Tang dynasty who wrote numerous poems about tea was an active exponent of the tea ceremony and of the culture of tea.

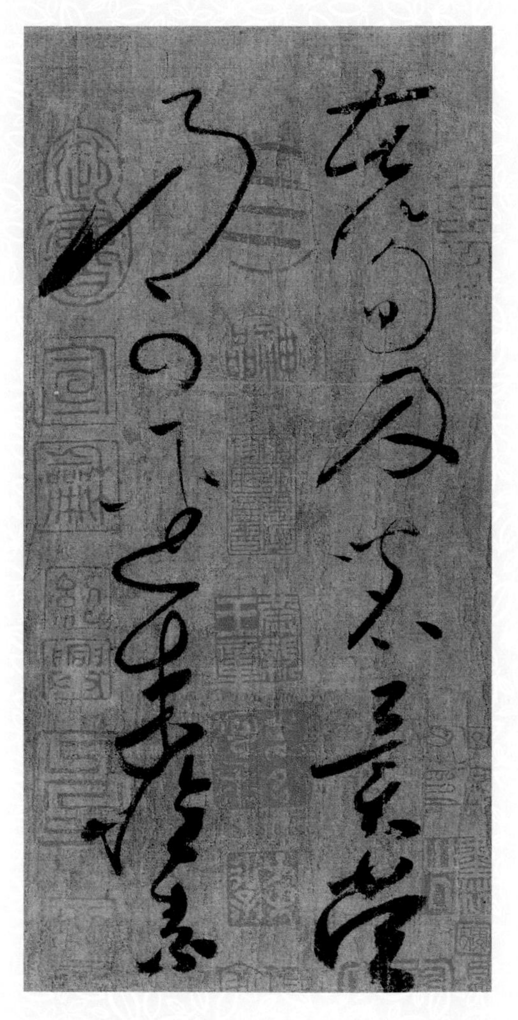

◀ A Request for Tea

This piece is written by Tang dynasty monk and calligrapher Huai Su (737 – ?). Now in the Shanghai Museum.

"Bamboo shoots and fresh tea leaves are both exceptionally fine, please send some directly. Respects, Huai Su."

A letter from Huai Su to a friend, the earliest known reliable piece of Buddhist calligraphy to do with tea.

The Buddhist tea ceremony gradually became the vogue during the Tang dynasty. Huai Su left home very young and in his own words mastered the three styles of grass script. He was known as "the mad monk." This relaxed piece of calligraphy has an elegant vigor and is a fine example of Huai Su's superb script.

Tasting tea beneath the bamboos, its fragrance finer than wine, and mundane thoughts vanish at once. The cicadas sound again in the trees, the sun has sunk in the west but my taste for tea is as strong as ever.

—*Qian Qi: A Tea Banquet for Zhao Ju*

Qian Qi (c.722 – 780), Tang dynasty poet.

▲ *Han Xizai's Night Banquet* (part)

Painted by Gu Hongzhong of the Five Dynasties, Palace Museum, Beijing.

Gu Hongzhong (c.910 – 980), Southern Tang artist. *Han Xizai's Night Banquet* depicts the scene at a night banquet given by the Southern Tang minister of the secretariat Han Xizai. In the picture the host sits on a couch and the guests sit or stand listening intently to a lady playing a Chinese musical instrument. From this painting it is possible to understand how at the time the tasting of tea was an important part of the banquets given by the nobility.

茶經卷上
唐竟陵陸羽鴻漸撰
一之源
茶者南方之嘉木也一尺二尺迺至數十尺其
巴山峽川有兩人合抱者伐而掇之其樹如瓜
蘆葉如梔子花如白薔薇實如栟櫚葉如丁香
根如胡桃瓜蘆木出廣州似茶至苦澀栟櫚蒲葵之屬其子似茶胡桃與茶根皆下
孕兆至瓦礫苗木上抽其字或從草或從木或從木幷從草
當作茶其字出開元文字者義從木當作搽其字出本草草木幷作荼其字出爾雅其名

When making tea and the water boils with a slight noise and bubbles like fishes eyes, that is known as "the first boil." When it boils at the edge of the pot with a curtain of pearl-like bubbles, that is known as "the second boil." When it boils in waves that is known as "the third boil." To continue boiling after the third boil is to over-boil and is unsuitable for tea-making.

—*Lu Yu: The Classic of Tea*

◄ *The Classic of Tea*

Printed in 780 and the first comparatively complete and comprehensive study of tea. Compiled by Lu Yu, regarded as the author of *The Classic of Tea* on the basis of the knowledge gained from ten years of research. The complete book contains more than 7,000 characters divided into three volumes (*juan*) and ten chapters.

One: Origin. This part tells the origin of tea trees and the appearance, names and quality of tea.

Two: Tools. This chapter describes the tools for picking and manufacture.

Three: Making. This chapter details the recommended procedures for the production of tea.

Four: Utensils. This chapter describes twenty eight items used in the brewing and drinking of tea.

Five: Boiling. This chapter describes the method of making tea.

Six: Drinking. This chapter details drinking customs and ways of drinking tea.

Seven: History. This chapter tells the stories of tea drinking and its efficacy as a medicine.

Eight: Growing Regions. This chapter describes places of production of famous teas and qualities of teas.

Nine: Simplify. This chapter lists those procedures that may be omitted and under what circumstances.

Ten: Pictorialize. This part provides an abbreviated version of the previous nine chapters.

▲ Gold Plated Tea Mill

This Tang dynasty gold plated tea mill was excavated at Famen Buddhist Temple in Xi'an, Shaanxi province, and is now in the Shaanxi Province Famen Temple Museum.

In the Tang dynasty frying tea (*jiancha*) was the major way of drinking it. The procedure was as follows:

1. Milling the tea. Taking the biscuit of tea and roasting it and when it had cooled, placing it in the groove of the mill so that the pressure of the wheel grinds the biscuit into tea ends.

2. "Frying" the tea by tipping the tea ends into a water pot and boiling while stirring with a spoon, then decanting it into a bowl for drinking.

Tea
Fragrant leaves,
tender buds
Admired by poets,
loved by monks
Tea mills carved from white jade,
tea baskets made from red muslin
Heated in the pot like yellow stamens,
poured in the bowl like yellowish flowers
Share it with the bright moon at depth of night,
summon the red dawn at break of day
Present and past washed away yet unwearied,
inebriated and praise it still.

—*Yuan Zhen: Poem of the Jewelled Tower*

Yuan Zhen (779 – 831), Tang dynasty poet, with Bai Juyi known together as "Yuan Bai."

This poem originally had the shape of a tower with the shortest line (one character) at the top and the longest (14 characters) at the bottom in the progression 1-4-6-8-10-12-14. The title *Poem of the Jewelled Tower* reflects this.

One bowl moistens the throat and two banish melancholy.
Three scour the stomach to inspire writings by the volume.
Four sweat life's troubles out by the pores.
Five lighten the limbs and six bring ecstasy.
Drink no more than seven and float forth on the breeze.

—*Lu Tong: Seven Bowls of Tea*

Lu Tong (795 – 835), Tang dynasty poet. He wrote the well known poem *Seven Bowls of Tea* after having received a gift of new tea from a friend. The poem describes his sensations after drinking the tea and vividly illustrates its benefits.

► *Lu Tong Making Tea*

Painting by Ding Yunpeng (1547 – 1628), now preserved in the Palace Museum, Beijing. Ding Yunpeng was skilled in figure painting, Buddhist images, and landscapes. The painting shows Lu Tong seated in a grove of banana trees, holding a circular fan and watching the stove upon which teas is being made.

▲ Pu'er Tea

One of China's historically famous teas.

Type: Black tea. Black is the third variety of Chinese tea and second only to green and red tea in annual production. It is highly fermented and derives its name from the fact that its leaves are glossy black or blackish brown in color.

Area: Yunnan province.

Characteristics: Made from the fresh leaves of the large leafed variety of the Yunnan tea tree. The loose leaf looks tightly rolled and gray/black in color. After infusion the water is a pale red and of the consistency of red wine.

The nature of tea is purity, drinking it causes one to forget the troubles of the world of dust (a Buddhist term for the mundane world). The flavor of tea alone approaches delicacy. It was originally grown in hill plantations. At my leisure I planted a few bushes here and there in the garden and to my surprise they grew vigorously, rather like an extra friend with whom one could talk.

—*Wei Yingwu*

Wei Yingwu (737 – 792), Tang dynasty poet.

Best Utensils: An earthenware teapot and a bowl with a lid both increase the strength and retain the flavor of Pu'er tea.

Varieties of Pu'er Tea:

Raw Pu'er: Kept for three to five years after natural fermentation before drinking.

Processed Pu'er: Artificially fermented and ready for drinking immediately.

With loose leaf Pu'er tea it is possible to compress the tea into bowl shaped cakes of tea (*tuo cha*), round flat biscuits of tea (*bing cha*), and rectangular or square tea bricks (*zhuan cha*).

靜思悟道

Song and Yuan Period

(960 – 1368)

The Song dynasty was the most flourishing period for tea drinking in China. The method of tea making developed from boiling or simmering to *diancha fa*, that is having milled the tea biscuit into tea ends, putting the tea ends in a tea bowl and then adding water and stirring with a spoon. From the court to the common people the tea drinking habit gradually extended to every corner of daily life.

This period also saw the beginning of the emphasis on the interest and aesthetics of tea-drinking. The environment, utensils and cultivated accomplishments of the tea drinker were all regarded as an important component of the art of tea drinking. More than 30 reasonably complete and representative books about tea were published. They recorded in detail the thriving production of tea during the period and described enquiries into the art of tea drinking.

◀ Calm thought realizes the way, patient toil fashions the vessel.

Tea encapsulates the spirit of hills and streams, it rids the heart of misery and melancholy, it revives the soul and calms the mind, but most people are unaware of these merits.

—*Zhao Ji: On Tea*

Zhao Ji (1082 – 1135) was Huizong, the Southern Song emperor and an accomplished calligrapher, artist, poet and literary figure. He was passionately fond of tea and made a profound study of it and wrote *On Tea* in the Daguan year (1107). The book has 20 sections and is a detailed description of tea growing areas, the picking and processing of tea, the methods of making it, its quality and the vogue for tea competitions.

► *Illustrated Guide to Tea Utensils* (*Cha Ju Tu Lu*)

Compiled in 1269 by Shen An Laoren (Old Man Shen An). The book illustrates and describes, a page at a time, the cultural characteristics of 12 different kinds of Song dynasty tea utensils for *diancha fa*, allowing us to know what they looked like.

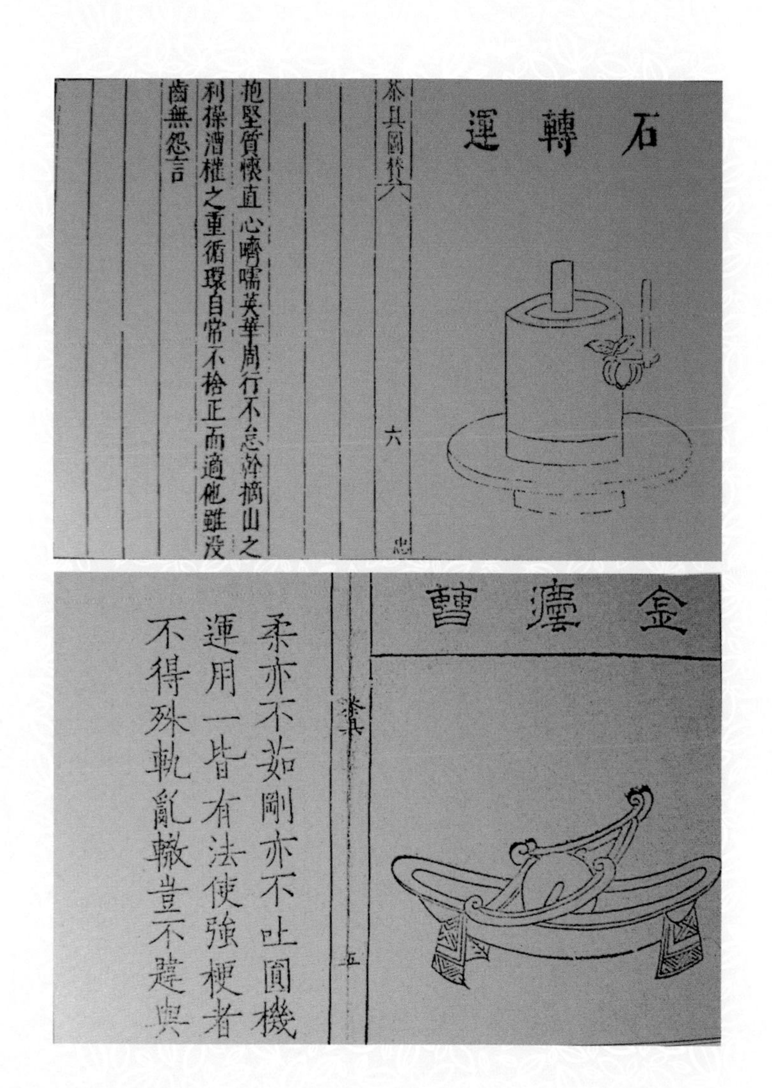

石轉運

茶具圖贊 六

抱堅質懷直心嚌嚅英華周行不怠斡摘山之利操漕權之重循環自常不捨正而適他雖没齒無怨言

金灋曹

茶具圖贊 五

柔亦不茹剛亦不吐圓機運用一皆有法使強梗者不得殊軌亂轍豈不韙與

If eyes are clear and body strong what harm lies in age?
If the rice is white and the tea sweet I shall feel no lack.

—*Lu You: On Joy*

Lu You (1125 – 1210), Song dynasty poet. It is said that Lu You wrote over 30,000 poems during his life, over 9,000 have been handed down of which more than 300 are about tea. Crowned first amongst historical poets.

◄ *Tea Fighting*

Ming dynasty wood block.

Diancha fa was the most important part of the process of tea drinking during the Song dynasty, that is stirring the tea with a spoon so that the tea tips and the water are evenly mixed into a milky paste. At this stage minute bubbles form a foam on the surface of the bowl, rather like white flowers, which adhere to the rim and stay there, this is called "biting the bowl."

With the rise of the Song practice of *diancha fa*, a form of social activity, *doucha*—tea fighting, a competition in judging the quality of tea gradually appeared. Victory or defeat in tea fighting depended upon: judgement of the merits and defects in the quality of the tea, the color of the tea and skill of *diancha fa*. This latter skill was crucial.

In tea fighting, the failure of foam to appear on the surface of the tea, or the failure of foam to bite the rim is known as "dispersal." Dispersal leaves a tide-mark on the rim. The loser is the one whose the tide-mark appears first.

The fewer guests when drinking tea the better. A crowd of guests is noisy and noise detracts from the elegance of the occasion. Drinking tea alone is serenity, with two guests is superior, with three or four is interesting, with five or six is extensive and with seven or eight is an imposition.

—*Cai Xiang: A Record of Tea*

Cai Xiang (1012 – 1067), Song dynasty calligrapher. His *A Record of Tea* or *Cha Lu* appeared in 1051. It contained approximately 800 characters in two volumes. The first volume mainly deals with the discrimination of the qualities of Song tea, and the methods for making it. The second volume is mainly a critique of the color and fragrance of tea leaves, the procedures for making tea, and the functions and use of tea utensils. This book is the next most well known specialist book on tea after Lu Yu's *The Classic of Tea*.

▲ Black Glaze Tea Bowl (*zhan*) with Rabbit's Hair Pattern

Song dynasty tea drinking utensil. So called because of the appearance in the black glaze of a network of striations resembling the fine hairs in the fur of a rabbit.

▲ *Serving Tea*

Tomb fresco, Liao dynasty (916 – 1125).

One person is seated on the ground milling tea and another kneels while vigorously fanning the fire to heat the water. A tall table at the left bears a set of a variety of tea utensils. The whole picture displays the procedure for taking tea from the milling of the tea to *diancha fa*.

One day, Su Shi and Sima Guang were tea-fighting for amusement. Su Shi's white tea won. To make it difficult for him Sima Guang then asked: "The whiter the tea the better, the blacker the ink the better; the heavier the tea the better, the lighter the ink the better; the newer the tea the better but the older the ink the better. Tea and ink are completely different. How can you like them both?"

Su Shi thought for a while and then replied: "Although there are enormous differences between tea and ink, perfect tea and perfect ink have one thing in common—fragrance."

—*Tu Long*

Tu Long (1543 – 1605), Ming dynasty literary figure and dramatist. Su Shi (1037 – 1101) was a Song dynasty literary figure, calligrapher and expert in the tea who wrote over 100 poems about tea and provided a detailed record of well-known Song teas, spring water, and tea-making and tasting. Sima Guang (1019 – 1086) was a Song politician, historian and literary figure. Ink (*mo*) refers to the sticks of ink used for writing and painting in ancient China.

Tea and bamboo shoots enshrine the flavor of Zen.
Pines and firs transmit true Buddhism.

—*Su Shi*

▶ *Fresco of* Diancha Fa

Tomb fresco, Liao dynasty (916 – 1125).

During the Song dynasty (960 – 1279) the major method of making tea developed from the Tang practice of frying tea (*jiancha fa*) to *diancha fa*. The attendant on the right of the fresco is adding water to the tea bowl with a water bottle and the one on the left is stirring with a silver stick. It is an extremely valuable historical pictorial example of *diancha fa* at the time.

At freezing night—a guest, I give tea for wine,
On the stove the water simmers above the leaping flame.
Beyond the window that same moon,
Now changed by the plum blossom.

—*Du Lei: At Freezing Night*

Du Lei (? – 1225), Song poet.

◀ *A Literary Party* (part)

By Zhao Ji. Now in the Palace Museum, Taibei.

The painting shows the splendor of a Song literary party. The bottom of the painting depicts the preparation of tea with a stove, tea bowls and other utensils, demonstrating the indispensability of tea at this kind of literary function at the time.

At first tea tastes bitter and then sweet. This is like a principle of life; bitter in the beginning, but in the end one can enjoy a little ease.

—*Zhu Xi: Writings*

Zhu Xi (1130 –1200), Song ideologist, philosopher, poet, educationist and literary figure. Zhu Xi's *Writings* or *Zhuzi Yulu* are a record of question and answer conversations with his students. They reflect his ideas.

▲ *Scences along the River during the Qing Ming Festival* (part)

Painted by the Southern Song painter Zhang Zeduan (1085 – 1145), now in the Palace Museum, Beijing.

This painting gives a glimpse of the flourishing teashop life of the Song and Yuan dynasties in the bustling Northern Song (960 – 1127) capital Kaifeng. There are more than 20 tea-houses in the painting. At the time tea-houses were not only for the leisurely drinking of tea but were places for doing business and meeting friends.

Ming Dynasty

(1368 – 1644)

By the Ming dynasty and with the unceasing improvement in the techniques of tea processing, the varieties of tea increased to nearly 100 to produce, one after another, green, black, dark green and red tea. At the same time the technique of adding boiling water (*paocha fa*) gradually replaced the technique of adding water several times (*diancha fa*) as the commonest means of making tea.

With the development of the art of tea-drinking into the Ming dynasty there was a greater search for peace and elegant seclusion as a mirror to the desirable culture of privacy of scholars and officials.

◀ Into autumn.

The following situations are suitable for the drinking of tea: a mood of leisurely content, the absence of anxiety; the presence of intimates of a similar cast of mind; whilst sitting in peace in a secluded spot; whilst reciting poetry; whilst exercising the brush in calligraphy and painting; whilst strolling through courtyards; on waking up; on waking the gods. Drinking tea should be accompanied by fruit and tid-bits, the table should be prepared, the surroundings should be elegant, there should be a meeting of minds between host and guest; there should be understanding and discrimination in the enjoyment of tea and there should be a boy to serve the tea and heat the water. The following situations are not suitable for the drinking of tea: where the tea has been ill-prepared; where the utensils are unsuitable; where there is friction between host and guest; official functions; feasts; when there is pressure of affairs, and when there is no interest in drinking tea or where the surroundings are chaotic and lacking in attraction.

—*Feng Kebin*

Feng Kebin, Ming dynasty tea expert.

▲ The Pourer and the To-and-Fro Method of Pouring Tea

Characteristics: Capacity greater than that of a teapot or a bowl with a lid, has one handle or two.

Material: Earthenware, porcelain, glass

Function: To achieve an even consistency in the strength of tea by using the pourer to pour tea that has been brewed in a pot or covered bowl into small tea bowls.

The to-and-fro method of pouring tea is employed to achieve an even consistency of color and aroma in the tea in the small cups by using the pourer to pour the tea backwards and forwards into the cups.

Taking two cups as an example, the procedure would be to fill the first one a third full, the second two thirds full and then to pour backwards and forwards between the two cups until both are seven parts full.

A young man who had suffered reverse after reverse sought help from the Zen master Shi Yuan. After hearing the young man's account of his suffering the master infused a bowl of tea with warm water and handed it to the young man saying: "Please drink a bowl of tea."

The young man took a mouthful and asked: "What tea is this? Why is there no fragrance?"

The master replied: "This is the very best Iron Buddha tea, how can there not be fragrance?" The master then told a disciple to bring a pot of boiling water. The master took another bowl, put tea leaves in it and added boiling water. Fragrance filled the air at once.

The young man raised the bowl to drink and the master said: "Slowly, slowly."

The master poured more boiling water into the bowl and the tea leaves rose and sank. He did this five times before the tea was properly made. At this point, fragrance filled the whole room.

The master then asked the young man: "The

same Iron Buddha tea, but why was the aroma completely different?"

The young man replied: "One bowl was filled with warm water and one bowl with boiling water, it was because the water was different."

The master said: "The temperature of the water is the key. When the tea was made with hot water the leaves floated lightly on the surface and didn't rise and sink, so how could they emit their fragrance? When boiling water is used again and again the leaves rise and sink over and over so that they can at last release the quiet seclusion of spring rain, the heat of the summer sun, the mellowness of the autumn wind and the crisp chill of winter frost."

When is life not like tea?

When a tea leaf meets boiling water it rises and sinks to release its fragrance, when life encounters the winds and rain of reverse and frustration it releases its glory.

—*Chinese Zen Buddhist Story*

Tea drinking is the highest form of the art of life. Idle chat comes second and drinking last.

—*Yuan Hongdao*

Yuan Hongdao (1568 – 1610), well known literary figure of Ming dyansty.

◀ *Making Tea* (part)

By the Ming artist Ding Yunpeng (1547 – 1628), and now in the Wuxi Museum.

The painting depicts numerous tea utensils. On the tea table there is an earthenware pot, a jar containing tea leaves, a bamboo strapped stove for making tea, tea bowls and saucers, an incense burner and a wicker basket. The bamboo strapped stove was much valued by the Ming literati.

It can be said that the contents of a bowl of tea are no simple thing. First there is the cultivation of a sprout of tea to make it grow, so that with constant irrigation and manure together with light, air, and water, it could be said that it grows to maturity by incorporating within it every power of the universe—how then could a tea bowl not contain all that exists? The Zen master Tou Zi maintained from start to finish that the universe was a bowl of tea, and that a bowl of tea was at the heart of the universe.

—*Ancient Chinese Zen Buddhist Story*

▶ Teapot from the Tomb of Wu Jing

Earthenware (*zisha*) teapot excavated from the tomb of the eunuch Wu Jing of the Jiajing period (1522 – 1566). It is the earliest reliably dated earthenware tea pot to have been found so far.

Zisha is an earthenware produced from the clay of Yixing. From the middle of the Ming dynasty onwards, *zisha* teapots became the nouveau riche of tea utensils and because of the esteem in which they were held by the literati and gentry, rapidly became an indispensable adjunct to the lives of the bookish classes. Even now the *zisha* teapot is still a well-loved form of tea utensil amongst Chinese tea drinkers.

Tea will emit its true fragrance when the time is ripe, neither early nor late. Too early and the fragrance will be insufficient, too late and the fragrance will disappear in the blink of an eye. Because everybody is different, the secret of this skill can only be slowly mastered by each individual mind.

—*Feng Kebin*

▲ *The Tea Party at Mt. Huishan*

Painted by Wen Zhengming (1470 – 1559), one of the most well-known calligraphers and artists of the mid-Ming dynasty. Ink and color on paper, now in the Palace Museum, Beijing.

This painting depicts the scene at one of the cultivated meetings most favored by Ming scholars, an assembly of three to five like-minded friends to drink tea in a secluded outdoor spot amid streams, stones, bamboo and pine.

Wings Waterfall

Tea is the soul and water is the mind. If tea encounters bad water it cannot display its grace and if water encounters bad tea it cannot show its quality.

—*Zhang Yuan: Record of Tea*

Zhang Yuan, Ming scholar, his *Record of Tea* or *Cha Lu* is based upon the methods of adding boiling water to loose tea. The book comprises about 1,500 characters altogether and is divided into 23 sections dealing with the picking of tea, recognising tea types, the making of tea, storage, controlling the flame and judging tea water.

◀ Good Water Makes Good Tea

Water suitable for tea should be clear, light, sweet and alive. "Clear" means that it should be translucent and without sediment. "Light" means that its mineral content should be low since minerals can affect the taste. "Sweet" means that the water should taste sweet. "Alive" means that it should be running water as the oxygen content of running water is high and helps the dispersal of the fragrance.

Remote from the clamor of this world of dust, in a tea hut surrounded by water amongst the trees in the depths of a hill-side wood, a close friend calls just after the Grain Rain and at the best time for the drinking of freshly picked tea. The new tea is made and drunk sitting face to face.

—*Wen Zhengming: Drinking Tea*

Grain Rain, one of the 24 Chinese solar terms. The period of the last ten days in April when plentiful rainfall benefits grain crops.

▶ *Drinking Tea* (part)

This painting by Wen Zhengming depicts mountains and streams, a thatched hut amongst pines and two people sitting face to face drinking tea. The painting is a vivid illustration of the remote and refined elegance that accompanied the cultivated practice of tea drinking.

▲ The Phoenix Bows Thrice

When hot water is added to tea leaves the pot containing the water should be held at a certain height and the boiling water poured on to the inner edge of the bowl in a rhythmic up and down motion repeated three times. This not only ensures consistency of density in the tea but also assists proper release of the taste of leaf.

This method of making tea has another significance in Chinese tea art where it is called "the phoenix bows thrice" representing the host greeting his guests with three bows as an expression of his courtesy and respect.

One day, the four friends Tang Bohu, Zhu Zhishan, Wen Zhengming and Zhou Wenbin went on an outing together. Having wined and dined heavily in Taishun they all felt sleepy.

Tang Bohu said: "I've heard that Taishun tea is excellent, why don't we each have a bowl to revive us?"

Zhu Zhishan said: "How can we have tea without poetry? Let's each make up one line of a poem on tea."

Tang Bohu then said: "Afternoon drowsy and all must sleep."

Zhu Zhishan added: "Drink green tea scented and sweet."

Wen Zhengming continued: "To tea one could add poetry too."

Zhou Wenbin finished: "The better to recite it all to you."

—*Ancient Chinese Tea Story*

Qing Dynasty

(1644 – 1911)

Following the rise in the vogue for tea drinking, tea acquired a close relationship with daily life and during the Ming and particularly the Qing dynasties many types of tea house were established. These tea houses became important centers of social activity, meeting the demands of all strata of society.

In the Qing dynasty the culture of tea made deep inroads amongst the general populace and local tea cultures with their own particular characteristics flourished, enriching the culture of tea in China as a whole.

◄ The finest flavor comes when hands are folded in silence.

True Dragon Well tea is cool to the taste, fragrant and sweet. It seems flavorless when swallowed but leaves a sense of great harmony which refreshes the mouth. This flavorless flavor is the true flavor.

—Lu Ciyun

▶ Xihu Longjing (Dragon Well) Tea

One of the teas with the greatest reputation in China.

Type: Green tea.

Area: The Dragon Well tea area of Lake Xihu near Hangzhou, Zhejiang province.

Characteristics: Smooth flat outer appearance, yellow green in color, tea water pure sweet and fresh. Normally, the tea picked before the Qingming Festival is known as "pre-Ming tea" and is of the best quality. That picked before the Grain Rain is known as "pre-Rain tea" and is of lesser quality.

Best utensil: A transparent glass.

This image shows the Longjing village which produces Dragon Well tea.

龍井問茶
從來佳茗似佳人

▲ The Etiquette of Tea

The Chinese custom of ceremonially making gifts of tea dates from the Tang dynasty in the 8th century when fine tea was distributed as a mark of regard and respect. The custom still continues today when new tea comes on the market and the Chinese send it to their close friends.

One day, the emperor Qianlong visited the Dragon Well tea area in disguise and took tea in a small tea-house, where the attendant, unaware of the emperor's identity, gave him the tea-pot and asked him to serve his entourage. Unwilling to divulge his identity the emperor poured tea for his followers. This frightened his entourage out of their wits and, in their anxiety they began to tap on the table with their fingers bent to show that they were "kneeling on both knees *ke tou* continuously" (*ke tou*—to knock one's head on the ground in obeisance).

This story later became wide-spread and "tapping the table in respect" became part of the etiquette of drinking tea even up to the present day.

When the host presents his guest with a tea bowl and pours water for him the guest slowly and rhythmically taps the table with the bent middle and fore fingers of the right hand in a gesture of respect to the host for having been served tea.

—*Ancient Chinese Tea Story*

In the autumn of the Bingwu year (1786) I took a tour to Wuyi, Manting peak and the Tianyou temple. The monks hastened to greet us with tea. The bowl was as small as a walnut and the pot as small as a grape-fruit and poured less than an ounce. One should not swallow it at once, one should savor its aroma and then try its taste and consider it slowly in the mouth as if chewing. As one would expect the fragrance assails the nostrils and there is a lingering sweetness on the tongue. Trying one cup and then one or two more has a calming effect on the mind and brings joy to the spirit. In the beginning I felt that although Dragon Well tea was clear its taste was thin and that although Yangxian tea was of quality its flavor was modest. There was very much the difference of character that there is between jade and crystal. Hence the fame of Wuyi (tea) is not to be shamed. Moreover it can be infused three times without its aroma being exhausted.

—*Yuan Mei: Recipes from the Sui Garden*

Yuan Mei (1716 – 1797), Qing dynasty essayist.

▲ Wuyi Cliff Tea

One of China's historical teas.

Type: Wulong (Oolong) tea.

Area: Mt. Wuyi in Fujian province.

Characteristics: Rope-twist outer appearance, mild in character and combining the clarity of flavor of green tea with the richness of red tea.

Best utensil: Earthenware tea-pot for making and white porcelain bowl for drinking.

▲ Warming the Pot

Warming the pot with boiling water before making tea has several advantages. 1. It prevents the temperature being lowered by the walls of the pot and affecting the flavor when the tea is actually being made. 2. When the tea leaves are placed in the warmed pot the heat in the walls of the pot bakes the fragrance out of the leaves.

When heating the pot first pour boiling water into the pot, replace the lid and then pour boiling water evenly over the outside of the pot to prevent the heat within the pot escaping, thus maintaining the internal temperature.

Tea cools through very small amounts of bitterness and sweetness. It clears the mind and revives the drowsy. It dispels anxiety, cools the liver and gall bladder, cleans the lungs and stomach, clears the eyes and quenches thirst.

—*Wang Shixiong*

Wang Shixiong (1806 – 1867), Qing dynasty medical specialist.

Before drinking tea, cleanse the mind and still anxiety.

—*The Emperor Qianlong*

The Emperor Qianlong (1711 – 1799) reigned 1735 – 1795. Qing dynasty emperor.

▶ Earthenware Tea Caddy

Used for storing tea. The body of this caddy bears a poem by Emperor Qianlong. Now it is in the Palace Museum, Beijing. The Qing dynasty emperor Qianlong always had a great fondness for tea. He closely followed the manufacture of tea utensils, enthusiastically wrote poems for tea huts and preserved the concepts and ideas of tea drinking for posterity. He was considered the most representative and knowledgeable of Qing dynasty experts on tea.

Generally tea should be stored at a low temperature to retard ageing, in dry conditions to prevent damp and mould and away from light to prevent deterioration.

Make tea to accompany proper speech, play the *qin* to know fine sound.

—*Ancient Chinese Tea Poem*

◀ The Four Cultivated Pursuits and Drinking of Tea

The four cultivated pursuits of playing the qin, chess, calligraphy, and painting were, like the drinking of tea, interests that were enthusiastically followed by the literary classes in the past. They possess the same qualities of relaxation, enjoyment, and leisurely tranquillity as the drinking of tea and are thus an accompaniment to tea drinking.

Modern

(1911 – 1949)

The growth of the tea culture was not only demonstrated by its increasingly close relationship with physical life but also by the fact that it was recognised as a valuable part of the traditions and inheritance of Chinese culture as a whole. Tea drinking was no longer a matter of merely quenching thirst, what was sought more and more was the spiritual satisfaction to be found in the ease and leisure that it represented.

◀ Sampling tea in a banana grove.

One day in the snow, I plucked a twig of winter plum on the outskirts of town and took it home. I slowly made a pot of tea and then as I sipped the tea I enjoyed the winter plum too.

—*Wu Changshuo: Inscription on the Painting Plum Blossom and Tea*

Wu Changshuo (1844 – 1927), calligrapher, painter and seal carver.

▲ Earthenware Teapot

Made from clay, only found at Yixing, this is one of the most characteristically Chinese forms of teapot.

Types: Smooth, with a plain body. Ornamented, where the body is modelled with the shapes of leaves, fruit, trees, animals, birds, fish and insects. Segmented, where the body is geometrically shaped with numerous vertical divisions.

Advantages: No loss of the original flavor when tea is made, no deterioration, no extraneous flavor from the internal wall of the pot, the ability to accept sudden changes of temperature with a slow transfer of heat and without burning the hands.

Tea should be drunk beneath a tiled roof and under paper windows, with clear spring water and green tea, with plain earthenware utensils and with two or three friends with whom a decade of dreams may be realised in half a day's leisure.

—*Zhou Zuoren: On Drinking Tea*

Zhou Zuoren (1885 – 1967), modern Chinese literary figure.

◀ Lidded Tea Bowl

A set of utensils consists of a small bowl, tea cup and lid. The small bowl symbolises earth, the tea cup mankind and the lid heaven. Hence the saying "a bowl of the three talents."

Material: Porcelain.

Advantages: Transmits and maintains heat well.

Best for making: Pu'er tea.

To have good tea to drink and to be able to drink it is a kind of "pure joy." However, to have the benefit of this "pure joy" one first needs time and then that special sense acquired through practise.

—*Lu Xun: On Drinking Tea*

Lu Xun (1881 – 1936), Chinese writer and thinker.

▶ Making Pu'er Tea

Utensils: Lidded bowl, pourer, small tea bowls.

Optimum temperature: Approx. 100 degrees centigrade.

Steps:

1. Warming the cup, that is heating the utensils with boiling water.

2. Moistening the tea, putting the tea leaves in the lidded bowl, covering them with boiling water and then quickly pouring the water away so as to waken the aroma of the tea and to rid it of any dust.

3. Pouring boiling water into the lidded bowl from a height, replacing the lid and letting it stand for a while, then straining it into the pourer and finally pouring it into the small tea bowls to be drunk.

夫復何求

Contemporary

(1949 – Present)

For the Chinese today, drinking tea still remains the favorite form of relaxation. In public or private, early or late, alone or together, drinking tea has become an indispensable part of Chinese life.

◀ A pot of tea, a beautiful view, a moment of contemplation, what more could one want?

Seven bowls for fragrance and one pot for true delight. A thousand Buddhist chants are not worth a drink of tea.

—Zhao Puchu: Drinking Tea

Zhao Puchu (1907 – 2000), contemporary Chinese author, poet, calligrapher and Buddhist.

▶ The Six Gentlemen of the Tea Ceremony

Tea Spoon: Small and long-handled with a round shallow scoop. Used for spooning tea from the tea caddy. Usually made of bamboo.

Tea Tweezers: Used when washing tea bowls and to extract tea dregs from the teapot. Usually made of bamboo.

Tea Needle: Long and pin-shaped. Used for clearing the internal sieve and spout of the pot to maintain the flow of tea. Usually made of bamboo or wood.

Tea Funnel: Used for putting in the mouth of a small teapot to guide the tea leaves into the pot and to prevent them from being scattered outside the pot.

Tea Scoop: For ladling tea leaves from the tea caddy into a teapot. Usually made of bamboo.

Container: Cylindrically shaped and used for holding tea spoons and tweezers etc. Made of bamboo or wood.

The merits of tea are internal, it is neither raucous nor irritating. With a small bowl of tea, not quite full, liquid and fragrantly clear, friendship flows in relaxation and conversation slowly unfolds and expands to illuminate the innermost soul of intimate friends.

—*Ai Xuan: The Nature of Tea*

Ai Xuan (1922 – 2001), contemporary Chinese author.

◀ Making Red Tea

Utensils: A lustrous white porcelain bowl that can set off the red of the tea.

Optimum temperature: 95 – 100 degrees centigrade, the high temperature fully releases the phenol in the tea.

Clear: Without any additions, the way Chinese people like it.

Added: With the addition of sugar and milk.

Addicted to tea as I am I still obstinately prefer tea picked between the Qingming Festival and the Grain Rain at the end of April. Tea banishes worry and dispels anxiety and is the best way of achieving harmony with nature.

—*Han Zuorong: Confessions of a Tea Addict*

Han Zuorong (1947 –), contemporary Chinese poet.

▲ Picking Xihu Longjing (Dragon Well) Tea

Dragon Well tea from Xihu is picked with particular care. Every spring the tea growers pick the new leaves in the order of the four qualities:

Lotus Heart: When the freshly opened buds of tea picked during the three days before the Qingming Festival resemble the heart of a lotus flower.

Flagstaff: The tea picked between the Qingming Festival and the Grain Rain is known as Pre-Grain Tea when the tea stalk grows a small leaf which looks like a flag on the staff of the stalk.

Sparrow Tongue: Picked at the Establishment of Summer (5 – 7 May). So called because the two leaves on a single stalk resemble a sparrow's tongue.

Stem Tea: Picked a month later.

▲ The Huxinting Tea House of the Chenghuang Temple in Shanghai

The two stories, the structure and the old fashioned internal décor of this building have a history that dates back over 150 years. It is famous as the first tea house of southern China.

A small rural tea house then becomes a "picture of the floating world" and when tea has been poured, no matter whether those at the same table are strangers or intimate friends, the chatter box is opened. Village news, world events, human joy and tragedy, official and unofficial histories, the big broadcast stories or rural affairs can all be exchanged over a bowl of tea.

—*Yang Yuyi: The Rural Tea House*

Yang Yuyi (1939 –), contemporary Chinese essayist.

Drinking tea requires leisure, with leisure comes calm and with calm established come the visions of tea and the clarity of mind that comes from the sipping of tea; and then a sense of calm unhurried joy that gradually wells up from the bottom of one's heart.

—*Ge Zhaoguang: On Zen and Tea*

Ge Zhaoguang (1950 –), contemporary Chinese scholar whose field of study covers the histories of religion and thought in ancient China.

► White Haired Silver Needle

Famous Chinese historical tea.

Type: White Tea.

Area: Fujian province, Fuding.

Characteristics: Needle shaped with fluffy white hairs, tea water pale yellow. This tea is only very slightly fermented and is unkneaded. Slightly more time is required to make it so as to allow the tea to brew properly.

Best utensil: Translucent glass or white porcelain.

Optimum temperature: Approx. 85 degrees centigrade.

The process of making tea puts me in mind of the stages of life, raw childhood, heady youth, grave middle age, nostalgic prime of life and an ever weakening old age that gradually loses its taste for life.

—*Lin Qingxuan: The Aroma of Tea*

◀ The Circumstances of Tea Drinking

Drinking tea is a spiritual pleasure and comprises both the "physical realm" and the "realm of the senses." The "physical realm" refers to the surroundings, which should be elegant and distinctive. The "realm of the senses" refers to the mood which should be unhurried and leisurely. Only a combination of the two will truly realise the spiritual pleasure that tea can bring.

Daily life rather resembles a cup of tea, most people's tea and utensils are very much the same, thereafter those who are good at making tea, make tea that tastes much more fragrant and those who are good at drinking tea acquire an even more exquisite piece of information; that life requires preparation, it is not a matter of expensive tea but of the mood in which tea is drunk.

—*Lin Qingxuan: The Aroma of Tea*

► The Way of Drinking Tea

Different kinds of tea have different characteristics and tea should be drunk according to the season and the physical characteristics of the drinker.

Green tea: Cooling, quenches thirst and alleviates fever. Drink during spring, summer and autumn.

Red tea: Warming, dispels cold and warms the stomach. Drink during the winter.

Wulong (Oolong) tea: Neither cold nor hot. Drink during the autumn.

Do not drink tea early in the morning on an empty stomach. Do not drink tea when taking medication. Tea should be drunk hot, do not drink it cold. Do not drink strong tea before going to bed.